ELECTRICITY

Chris Ollerenshaw and Pat Triggs
Photographs by Jon Allyn

Contents

Words that appear in the glossary are printed in **boldface** type the first time they occur in the text.

Gareth Stevens Publishing
MILWAUKEE

What's in my toy box?

Do you have any toys that run on batteries? There are quite a few pictured here. Some came from stores. Some were made at home. Some toys need only one battery to make them work. Some toys need several batteries.

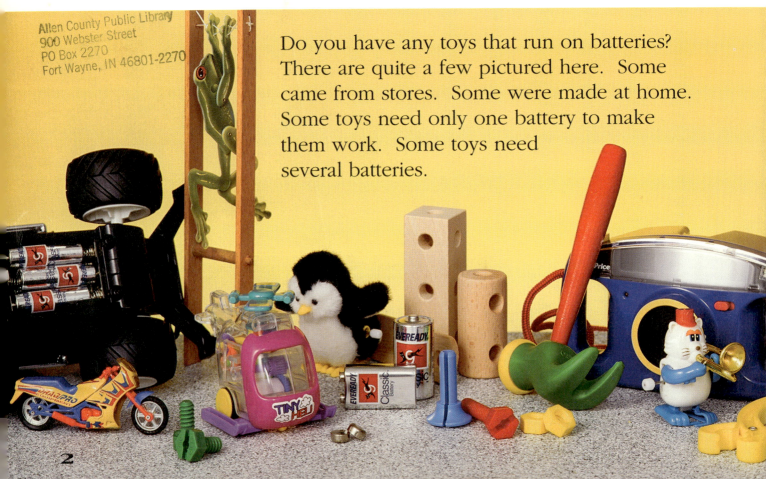

These items have batteries inside them. What items in your home operate on batteries?

Sometimes when a toy or a flashlight doesn't work any longer, people might say, "Maybe the battery is *dead*." What do you think that means?

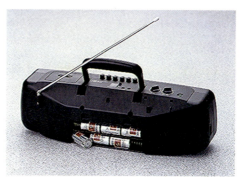

This cassette player runs on batteries. But it will also work if you plug it in. It can operate either on batteries or by plugging it into an outlet.

Plugging in

Many things that we use every day are plugged into outlets to make them work. Can you think of some?

We use these items so often that we need to be near a supply of the type of energy that makes them work.

What makes them work? Electricity!

Homes, schools, offices, and factories all use electricity that is supplied by the electric power company.

But wires from the power company do not go everywhere. We use batteries where there is no supply of electricity. Batteries store electricity so we can carry it around with us. Batteries are also safer to use than electricity from an outlet. Electricity from an outlet is very powerful and could kill you. A battery's power is not so strong.

What's the problem?

Sometimes toys that use batteries stop working. This model lighthouse has stopped working. Its light won't come on. How can you find out why the light isn't working?

The way to discover why something has stopped working is to understand how it works in the first place. You can find this out, step by step, by taking the lighthouse or a similar toy apart.

Ask an adult to help you take the toy apart. After you have it apart, you can see all of its parts, including the electrical ones. Then have the adult help you discover why the toy has stopped working.

Start with a battery and a bulb. By touching the bulb to the battery, see if you can get the bulb to light.

Which part of the battery did you touch with the bulb to make the bulb light?

Making connections

Next, use two pieces of wire to make the bulb light without having the bulb touch the battery directly. Where did you connect the wires to the battery? Where did you connect the wires to the bulb?

The battery, bulb, and wires are all made of different materials. These materials include metal, plastic, and glass. Which of these materials is allowing the electricity to flow? Which parts need to touch in order to make the bulb light? What are they made of?

Use your ideas about what allows electricity to flow to perform the next step. This time use the battery, the bulb, the two wires, and a bulb holder. You will also need a screwdriver.

Using the bulb holder

Connect the wires to the bulb holder. Using a bulb holder is very practical. Remember how awkward it was when you were trying to get the bulb to light by using your hands?

The bulb holder has been designed to hold the bulb firmly. This feature allows electricity to flow smoothly.

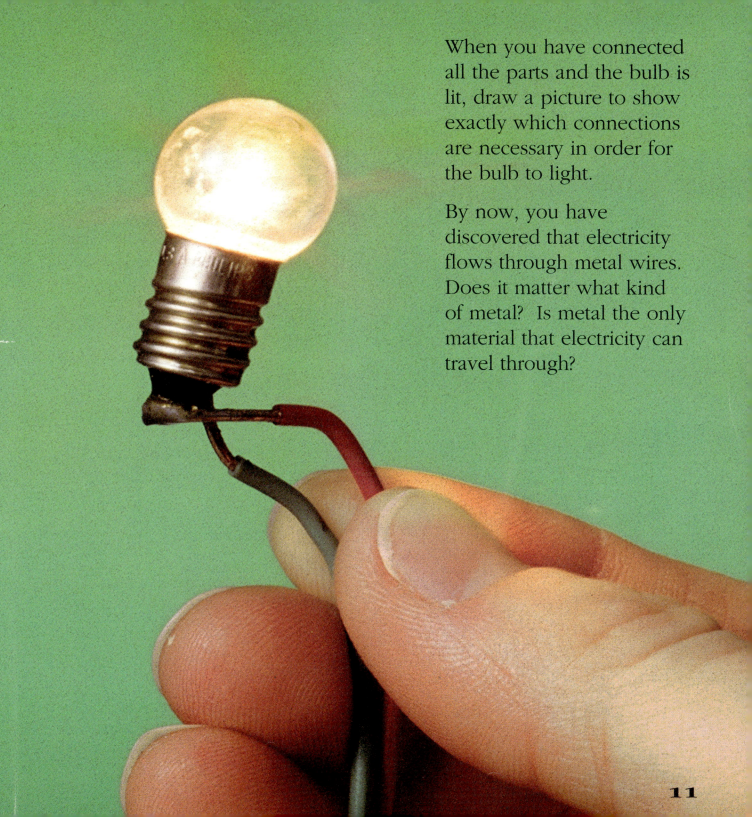

When you have connected all the parts and the bulb is lit, draw a picture to show exactly which connections are necessary in order for the bulb to light.

By now, you have discovered that electricity flows through metal wires. Does it matter what kind of metal? Is metal the only material that electricity can travel through?

Bridging the gap

There are certain materials that do not **conduct**, or allow the flow of, electricity. Collect various materials and guess which of them will conduct electricity.

Next, disconnect one of the wires from the battery and the bulb holder. Cut the wire in half (picture 1), but *never* cut through electrical wires while they are connected.

Have an adult help you strip the plastic coating from both ends of the wire where the cut was made. Then reconnect one half of the wire to the battery and the other half to the bulb holder (picture 2). Touch the two ends of the wires together momentarily to complete the **circuit**.

Use the different items you have collected to try to bridge the gap between the wires. Which item allows electricity to pass through it?

☀ YES ✓	💡 NO ✗
📎 paperclip	🎀 rubber band

The materials that allow electricity to pass through them are called **conductors**. The materials that stop the flow of electricity are called **insulators**.

How good were your guesses?

Before electricity

Homes, schools, and businesses did not always have electricity. It was not until the early 1900s that electricity became common. Imagine what life was like before then. The objects pictured were all used before the time of electricity.

Danger!

Electricity is very useful, but it must be used safely. Like metal wires, humans and animals conduct electricity. Make sure never to put anything, except a properly insulated plug, into an electrical outlet. Insulation materials are wrapped around objects that carry electricity so that the electricity and living things can be kept safely apart.

Safe connections

Have you ever seen someone start a car using jumper cables? Sometimes there is not enough power in a car's battery to start the engine. So jumper cables are used to connect the "dead" battery of one car to the working battery of another. (*Never* touch a car battery. You could get a shock or burns.)

The connections are made with strong crocodile clips that are made of metal. From looking at the clips, it's not hard to guess why they are called that. Can you guess what the handles are made of, and why?

The electrical plugs on appliances are designed to make safe connections. Many modern appliances have safety plugs like the one pictured below. This plug contains a feature that will shut the appliance off if the appliance is accidentally dropped into water.

Making a circuit

Next, gather all the materials that you found to be good conductors. Join the conductors together to make a long chain and see if the bulb will light. You will need to make sure that all the conductors are touching. Electricity can't cross gaps. It needs an unbroken path to flow along.

An unbroken path to and from a source of electricity is called a complete circuit. A circuit doesn't necessarily have to be in the form of a circle. Look at the shape of the circuit you have made. You can trace the path of an electric current around a complete circuit with your finger. Start at any point and follow the path through one conductor to the next, including the battery, and you will come back to the starting point.

How many objects were you able to include in your chain of conductors? As you added more objects, was it harder to keep the bulb lit or stop it from flickering? The more objects you added, the more hands you needed to hold them all in place!

Can you think of a way of making good connections all along your circuit? What materials would you use?

Controlling the flow

If all the connections in your circuit are good, the bulb will stay lit. When the bulb is lit, it is using the electricity supplied by the battery. When the battery stops moving electricity, the light goes out, and the battery is "dead."

The electricity that we use doesn't just happen out of thin air. Most of the electricity we use comes from either **generators** or batteries. Electricity that is in your home is generated in power stations and carried along cables until it reaches your home.

Houses, schools and other buildings

cables (these are often buried underground)

Power station

Electricity is expensive. It is paid for when we buy batteries or pay the electric bill. It should be used only when it is needed, otherwise it is wasted. So there has to be a way for us to turn it on and off.

On-off

To stop the flow of electricity, the connections must be broken. To break the connections, you could disconnect a wire, loosen a bulb, pull out a plug, or turn off a switch. A switch allows you to easily control the flow of electricity. There are many different kinds of switches.

But all switches do one thing. They open and close circuits in a way that is safe for the person using them.

How many different kinds of switches can you find in the world around you?

You know the circuit you have just made is working. How would you use what you've now learned to determine whether the parts of the lighthouse or another broken toy's circuit were working or not? You would need to check the battery, the bulb, the connections, and the switch.

Looking for clues

You could test the battery first. To do this, you need to take the battery out of a circuit that you know is working and replace it with the lighthouse or another broken toy's battery. Does the bulb light?

If the bulb doesn't light, it could mean that the toy's battery is "dead." You could try putting a new battery into the toy.

But what if the bulb still doesn't light? Then you would need to do further testing because sometimes more than one thing can go wrong at once.

To find out whether the bulb is the problem, you need to take the bulb out of your working circuit and replace it with the broken toy's bulb. If the bulb lights, it is fine and can go back into the broken toy's circuit.

Finding the answer

Suppose the battery and the bulb are both working properly, but the toy still doesn't light.

Then you need to make sure all the connections are good by checking that the wires are properly connected to the

battery and the bulb holder. In addition, make sure the bulb is screwed in tightly.

To test the switch, take it out and join the wires together. If the bulb lights, this means the switch isn't working properly.

Take some time to make your own model lighthouse, repair a broken flashlight, or put some lights into a dollhouse.

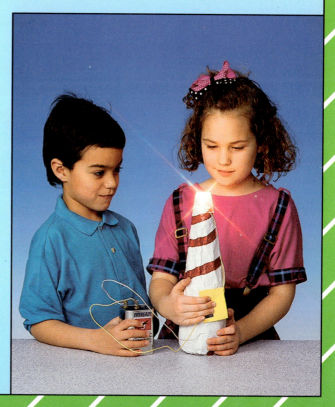

Making a model robot

You can use what you have discovered about electricity to build a robot frog. Plans for the robot are on pages 30-31.

Before you design the circuit that will make the eyes light, have an adult help you find out how to make two bulbs light at once.

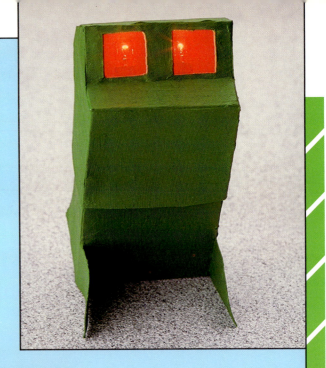

Can you find two different ways to make both bulbs light together? Can you think of a reason why the bulbs in one circuit will be brighter than the ones in the other? Which circuit will you choose to fit inside the robot you have made? Why is it better?

For a free color catalog describing Gareth Stevens' list of high-quality books, call 1-800-542-2595 (USA) or 1-800-461-9120 (Canada). Gareth Stevens' Fax: 414-225-0377.

The publisher would like to thank Richard E. Haney, Professor Emeritus of Curriculum and Instruction (Science Education) at the University of Wisconsin-Milwaukee for his assistance with the accuracy of the text; the Milwaukee County Historical Society, Milwaukee, Wisconsin, for historical artifacts; Cudahy News & Hobby Center, Cudahy, Wisconsin, for donated toys; and models Katie (cover), David, Jessie, and Nikki.

Library of Congress Cataloging-in-Publication Data

Ollerenshaw, Chris.
    Electricity / Chris Ollerenshaw and Pat Triggs. -- North American ed.
     p. cm. -- (Toy box science)
    Includes index.
    ISBN 0-8368-1119-4
    1. Electric power--Juvenile literature. 2. Electric apparatus and appliances--
Juvenile literature. [1. Electric apparatus and appliances. 2. Electricity.]
I. Triggs, Pat. II. Title. III. Series: Ollerenshaw, Chris. Toy box science.
TK148.045 1994
621.31'042--dc20
                                 94-4883

North American edition first published in 1994 by
Gareth Stevens Publishing
1555 North RiverCenter Drive, Suite 201
Milwaukee, Wisconsin 53212 USA

First published in 1991 by A & C Black (Publishers) Ltd., London. Original text © 1991 by Chris Ollerenshaw and Pat Triggs. All photographs © 1994 by Jon Allyn, Creative Photographer. Additional end matter © 1994 by Gareth Stevens, Inc. Illustrations by David Ollerenshaw (model plans) and Dennis Tinkler. Designed by Michael Leaman.

Series editor: Barbara J. Behm
Cover design: Karen Knutson

Printed in the United States of America

1 2 3 4 5 6 7 8 9 99 98 97 96 95 94

At this time, Gareth Stevens, Inc., does not use 100 percent recycled paper, although the paper used in our books does contain about 30 percent recycled fiber. This decision was made after a careful study of current recycling procedures revealed their dubious environmental benefits. We will continue to explore recycling options.

Robofrog Pattern

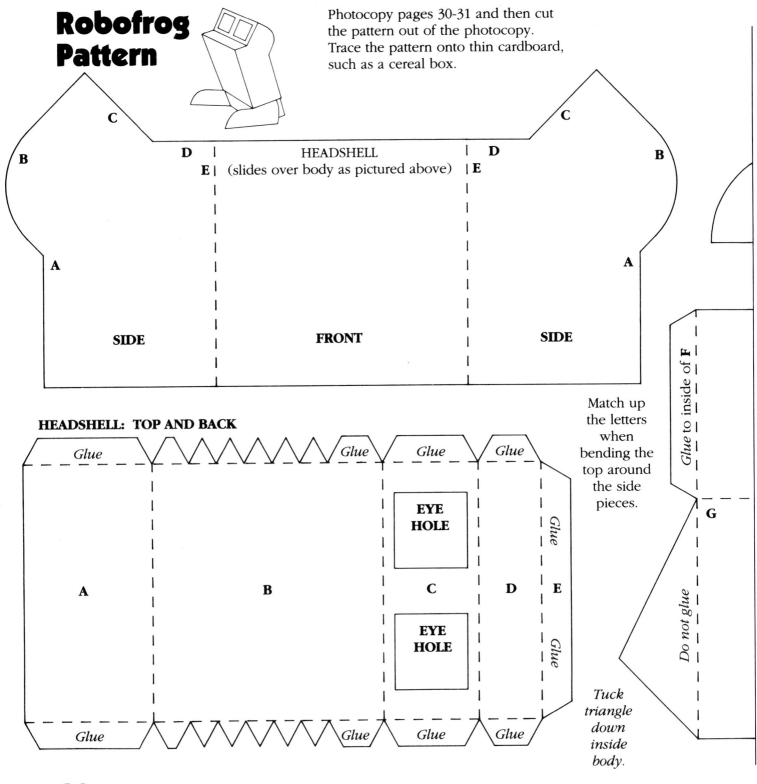

Photocopy pages 30-31 and then cut
the pattern out of the photocopy.
Trace the pattern onto thin cardboard,
such as a cereal box.

C

B

D

C

D

B

HEADSHELL
(slides over body as pictured above)

E

E

A

A

SIDE

FRONT

SIDE

Match up
the letters
when
bending the
top around
the side
pieces.

HEADSHELL: TOP AND BACK

Glue

Glue

Glue

Glue

A

B

EYE
HOLE

C

D

E

Glue

Glue

EYE
HOLE

Glue

Glue

Glue

Glue

Glue

Glue to inside of **F**

G

Do not glue

*Tuck
triangle
down
inside
body.*

CUT along all solid lines. **FOLD** along all dotted lines. **GLUE** the body together. Stick legs on as shown in illustration below. The bottom front edge of body should rest on the table. Make the headshell separately. It slides on and off the body like a lid.

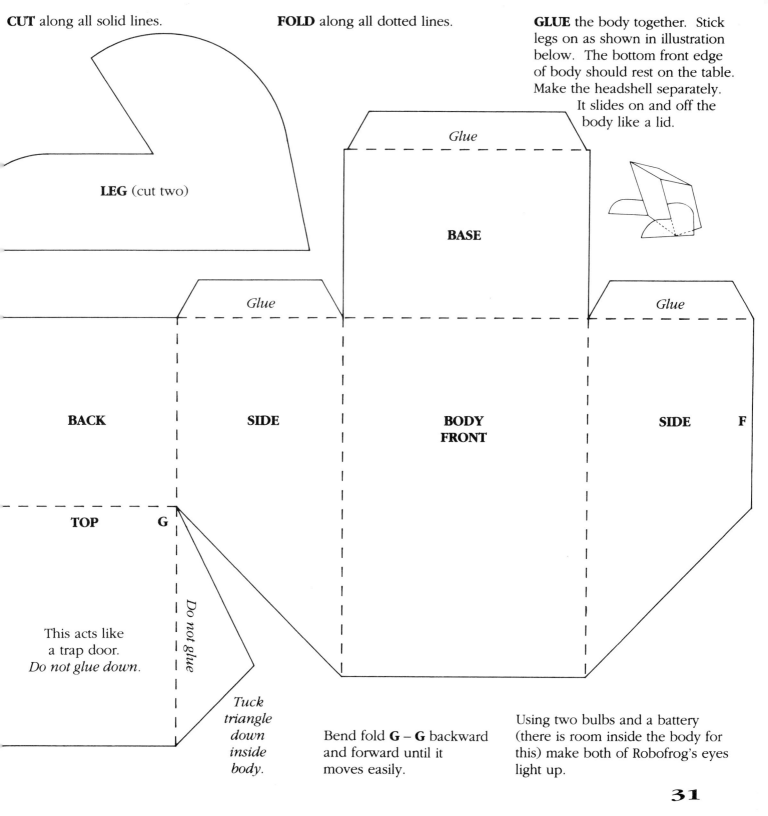

LEG (cut two)

Glue

BASE

Glue

Glue

BACK

SIDE

BODY FRONT

SIDE

F

TOP   G

This acts like a trap door. *Do not glue down.*

Do not glue

Tuck triangle down inside body.

Bend fold **G – G** backward and forward until it moves easily.

Using two bulbs and a battery (there is room inside the body for this) make both of Robofrog's eyes light up.

31

Glossary

circuit: a route that returns from where it started

conduct: to allow the flow of electricity, heat, or other forms of energy

conductors: materials that provide a smooth flow of electricity, heat, or other forms of energy

generators: machines that change mechanical energy into electrical energy

insulators: materials that cover an object to slow or prevent the flow of electricity, heat, or other forms of energy

Books and Videos

How Did We Find Out about Electricity? Isaac Asimov (Walker and Co.)

How Electricity is Made. C.W. Boltz (Facts on File)

Simple Science Projects. Projects with Electricity. (Gareth Stevens)

Science Discovery for Children. (video)

Index